Disaster and Climate Change Preparedness in American Sāmoa

This project was funded by the American Psychological Foundation Visionary Grant, with support from the Pacific Regional Integrated Sciences and Assessments (Pacific RISA) program and the East-West Center.

Kati Corlew, PhD, is project principal investigator for this project, "Relating the Psychological Recovery from Recent Disasters to Climate Change Risk Perception and Preparedness in Hawai'i and American Sāmoa." She is an assistant professor of Psychology at the University of Maine at Augusta. She can be reached at kate.corlew@maine.edu.

For a free electronic file, available for download, and to learn more about the Pacific RISA project, visit www.PacificRISA.org.

The handbook is also available at EastWestCenter.org/Publications.

For permissions requests contact EWCBooks@EastWestCenter.org.

Disaster and Climate Change Preparedness in American Sāmoa
ISBN 978-0-86638-258-8 (print) and 978-0-86638-259-5 (electronic)

Photograph sources in this publication:
Cover by Greg McFall
Pages 1, 3, 6, 13, and 17 by LT Charlene Felkley
Page 15 by Kati Corlew, PhD
Page 23 and 24 by Krista Jaspers

Overview

American Sāmoa is home to some of the most beautiful ecosystems on earth. But if you've lived here long, you know that American Sāmoa (like everywhere else) is vulnerable to natural and man-made disasters. Vulnerabilities include such risks as drought, wildfire, heavy storms and flooding, mudslides, erosion, tsunami, and earthquakes, among others.

Some of these hazards will be exacerbated in the coming years by changes to the climate. For this reason, disaster and climate change preparedness can go hand in hand. Many disaster preparedness actions will make families, business, and communities better prepared for climate change as well.

The purpose of this project is to better understand American Sāmoa's relationship with natural hazards and to help American Sāmoa citizens and professionals prepare for future events. This project connected with American Sāmoa community members about their experiences with hazard events in three different ways:

1) An online survey
2) Interviews with community members and professionals
3) A preparedness workshop in Pago Pago

This booklet includes information about natural hazards and vulnerabilities to disaster in American Sāmoa, stories from project participants about their experiences, and a guide to disaster and climate change preparedness.

Hazards versus disasters

What's the difference?

A **hazard** is a threat of an event (natural, man-made, technological) that could be harmful to people or the environment.

A **disaster** is a natural, man-made, or technological hazard event that results in physical damage, destruction, loss of life, etc.

> Living in a flood zone is a *hazard*.
> The Japanese tsunami was a *disaster*.

A disaster can sometimes set off **secondary disasters**, or even create a series of cascading disasters, like a domino effect. They can include natural and man-made hazards.

> In 2011, an *earthquake* of the coast of Tohoku, Japan caused a *tsunami* that *destabilized* a nuclear reactor...

The International Disaster Database (EM-DAT) uses the following criteria as a guideline for defining a disaster:

- 10 or more people reported killed and/or
- 100 or more people reported affected and/or
- Call for international assistance/declaration of a state of emergency

Recognizing hazards in American Sāmoa

American Sāmoa features many diverse landscapes all on one island and also has a rainy and a dry season. This variability means that American Sāmoa is home to a wide variety of weather-related and non-weather related hazards that interact with each other to create many types of secondary hazards in different places around the island.

American Sāmoa has experienced a number of **cyclones** and **tropical storms**, which may increase the occurrence of **flash floods** and **rock slides**. Close to shore, there is a danger of **rip tides** that must be considered. Additionally, American Sāmoa and the surrounding areas in the Pacific are vulnerable to **earthquakes** that may or may not produce a **tsunami**.

Natural hazards are hazards to communities.

- Natural hazards can damage infrastructure, like roads and bridges, or electricity and water supplies
- Natural hazards can interrupt community functions – from family functions to work and school to shipping and societal organization
- Natural hazards can cause injury and loss of life

Climate change increases hazards and disasters

Climate change is already affecting American Sāmoa, the Pacific Islands region, and the world. Though the severity and timing of the changes cannot be precisely predicted, scientists from the Pacific and around the world know many changes will happen.

In American Sāmoa, many species are at risk. **Endangered species**, already very vulnerable, are at increased risk.

Because of the threat to island and ocean species and ecosystems, there is a threat to **spiritual and cultural** activities.

Weather patterns will become more extreme. For example, there may be increased drought AND increased storm severity, which could increase impacts like flood, erosion, and rock slides.

Ocean acidification, erosion, and other threats may damage the **coast and near-shore environments**, creating unstable coastlines and reducing the natural storm buffer.

Remember, risks to the environment ARE risks to the community:

They threaten livelihoods, food security, and infrastructure.

Emergency Management Cycle

Disaster and emergency managers think about disasters as occurring in a cycle of stages, called the Emergency Management Cycle. Before the event is a time for **preparedness**. Immediately after the disaster strikes is the **response**. For the weeks, months, or even years after a disaster is the period of **recovery**. As the community becomes stable again, **mitigation** activities increase the community's ability to respond and recover from future disasters. This leads once again into **preparedness**.

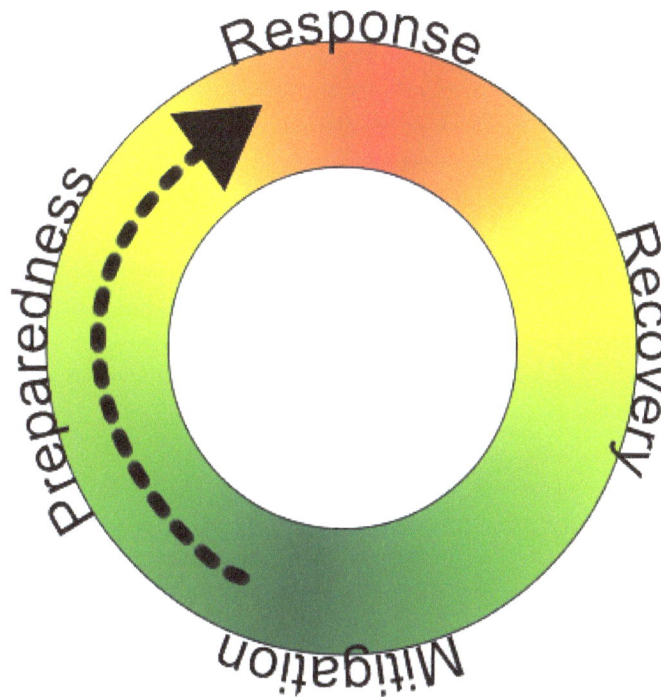

Image Source: http://en.wikipedia.org/wiki/File:Em_cycle.png

Experiencing Disaster

What happens psychologically before, during, and after a disaster occurs?

Many psychological experiences occur throughout the emergency management cycle: before, during, and after a disaster event. Some of these experiences are negative. Some are positive. Some are surprising, and though they are common, many people do not expect them.

Before the disaster event:

- Self-efficacy, or readiness – this can be increased by spending time preparing a disaster kit, making a plan, and learning about hazards
- Validating – when people know disasters are coming, they will often 'check in' with their loved ones to see if everyone else is making the same decisions they are, e.g., Are you boarding up your house? Are you evacuating?

During the disaster event:

- Proactive – when a disaster is occurring, people will take what time and opportunities they can to mitigate the impacts of the disaster, e.g., Grab your wallet or your shoes before evacuating from an unsafe location; close windows in a fire
- Reactive – when a disaster is occurring, people will often find themselves responding to events without conscious thought or reasoning; when events are happening quickly, there may simply be no time

Immediately after the disaster event:

- In the hours and days following a disaster event, people may experience anxiety, psychological trauma, and increased sense of risk; they will often find themselves expecting another disaster at any moment
- In the hours and days following a disaster event, people will often exhibit higher rates of helping others; mutual care for family, friends, and strangers; it is also common to see family and community mobilization as people help each other to recover from the disaster event

Ongoing recovery:

- In the weeks, months, and even years following a disaster, people may experience stress, sensitivity, intense feelings, and an on-going increased sense of risk to future disasters
- In the weeks, months, and even years following a disaster, people may also experience increases in self-efficacy, adaptation, and preparedness for future disasters. This is called **post-traumatic growth** and is associated with people learning what they are capable of doing during a time of extreme stress.

Heroic actions may occur at any point in this cycle. They may be proactive or reactive. They may be momentary or they may last for hours, days, or even years.

Survey results

What did people have to say about their disaster experiences? How prepared were they for future disasters?

The survey was conducted in American Sāmoa and Maui. Answers from 33 people were included in the final analysis. 14 (42%) people were from Maui, and 19 (58%) from American Sāmoa. Participants had both a personal and a professional interest in disasters. 17 (52%) were women, and 15 (46%) were men.

American Sāmoa participants had experienced:
- 2009 Tsunami, 14 (79%)
- Hurricanes, 12 (63%)
- Flooding, 3 (16%)
- Rip Tide, 1 (5%)

Experiences after the disaster (all participants):
- 7 (21%) *frequently* think or dream about their disaster experience
- 12 (36%) *frequently* discuss their disaster experience
- 21 (64%) are *very concerned* about future disasters
- 29 (88%) are now *more prepared* for future disasters

Preparedness for future disasters (all participants):
- 17 (52%) have a disaster kit at home
- 14 (42%) have some resources prepared, but do not have a disaster kit at home
- 25 (76%) have an emergency plan at home
- 24 (73%) have an emergency plan at work, but only 17 (52%) *practice* their emergency plan at work

Disaster stories from American Sāmoa

Community members and professionals in American Sāmoa participated in interviews to talk about their experiences with disaster and climate change. They discussed experiences with the 2009 Tsunami, with hurricanes, and with flooding caused by storms. They discussed disaster warnings that have happened since their disaster experience.

Disaster preparedness is a spectrum – from unprepared to very prepared. It is impossible to be 100% prepared for a disaster. No one can be ready for everything. However, the interview participants shared their stories and what they learned about how to be more prepared should disaster strike again.

Educate yourself

Before a disaster occurs, it is important to know what might happen, and what to do. When the 2009 tsunami hit, one woman was told to get to higher ground, and quickly.

"I got onto the main road and at this point, I didn't know how far inland I needed to go or how high up I needed to go, like where would be a 'safe area.' So all I could think of was, 'Okay I will go up Aoloau' which is the big mountain on the west side. So we went up that."

American Sāmoa now has evacuation signs to show where is a safe area in case of tsunami. But when the tsunami hit, this woman was confused and scared and did the best she could – she went as high as she could so that she knew she and her passengers would be safe.

One man told the story of a tsunami warning that happened overnight in 2011. The new tsunami warning sirens went off to signal there was a tsunami watch, but this man was new to American Sāmoa, and realized he did not know what he was supposed to do when he heard the siren.

> "My immediate thought was like "Oh it's a not a big deal," but then I started to think of the "what if and what if" and getting a little – it was keeping me up a little bit. So I got out of the bed and noticed that my neighbor's light was on and she was – I don't have a TV so or a radio inside my house, so I don't have really ways of tracking things live."

He went to his neighbor's house after realizing he did not have a way to receive official communications about evacuation. His neighbor, who was keeping track of the warning information, assured him they were high enough above sea level to be safe even if a tsunami did come.

One interview participant told the story of the Old Man of Amanave Village, who had received earthquake and tsunami training years before the 2009 Tsunami. They learned about how to tell if an earthquake was strong enough that it might generate a tsunami.

"When an earthquake is so strong that you get up and you are shaking and you fall down again, try to get up, things would be falling, all the typical signs... And then, one thing for sure, if you see the ocean receding, that's the sign that a tsunami is coming. I don't care if you felt the earthquake or not, if you see the ocean moving, going back, you know a tsunami is on the way."

In 2009, when the mayor felt the earthquake and saw the signs for the tsunami, he got one of his kids to drive his truck up and down the village, yelling for everyone to go up the mountain because the tsunami was coming.

"It was an example of somebody who was ready... He thought back on all the trainings... he just grabbed the bullhorn and went for it. That man saved his whole village."

Look for the signs

When people are prepared for disaster, they can look for the signs and respond accordingly. When a hurricane is coming, people may know days in advance and be able to prepare their houses and families for the storm. When an earthquake, tsunami, or flash flood from a rainstorm happens, there may be very little time to prepare a response. That is why it is good to be prepared ahead of time. That is also why it is good to pay attention to the signs. The mayor of Amanave village felt the earthquake and saw the ocean receding. But that day, not everybody experienced those signs.

> "What happened was I lived on the west side at the time, and I have a car and I always carpooled... and when we picked her up, she stops, we got her in the car, she is like, 'Man, did you feel that earthquake? It knocked me down.' And we were like, 'No, we are in the car, we didn't feel anything.'"

Many people who were driving did not feel the earthquake when it happened. One woman said she was in a bus so they did not feel the earthquake. She saw rocks on the road and wondered why they were there because there was no rain. She saw people driving the other way honking their horns and flashing their lights. Finally, someone stopped the bus and told them to run to higher ground. They did not feel the earthquake or see the ocean recede. They saw only secondary signs.

It is also important to watch for signs because it takes time for warning systems to mobilize. At the time, American Sāmoa did not have a siren system in operation, and only the Pacific Tsunami Warning Center could issue official warnings. Because the earthquake was so close, there was no time for an official warning to be issued. The Weather Service Office of American Sāmoa got on the handheld radio for the emergency alert system communication and issued a warning. Unfortunately, because there was so little time between the earthquake and the tsunami, the warning did not reach all communications. One radio station reported the tsunami as it was reaching the broadcast center in Pago Pago.

The warning systems in American Sāmoa have been updated extensively since the 2009 Tsunami, and will continue to improve in the future. However, it is important for community members to realize that they can prepare as well, so that when disasters occur quickly or if they damage communication systems, the communities can respond.

Care for yourself and each other

During and after a disaster, it is very common to see people act quickly and without thought. One woman told the story of running up the mountain when the tsunami was coming. She only remembers flashes – the sounds, the smells, her feet getting wet.

Another woman, in panic mode while she drove up to higher ground, stopped to pick up anyone who could fit in her car. They yelled out the windows to warn everyone to get to higher ground. People helped each other as they could, even in a state of panic. This help continued after the disaster as well.

All around Tutuila, people gathered food, water, clothing, and supplies for others who were impacted by the tsunami. Communities came together to help clear out buildings, to clean up villages, and to support each other through grief.

Though many people mentioned that American Samoans will often not talk about these experiences, it is important to realize that trauma tends to stay with people. They may recover quietly, but recovery is a process that happens over time.

One woman said, "Don't try to be lonely." She said that in the weeks and months after the tsunami, it was most important to be with family and friends. The support of loved ones helped her recover.

Preparing for disaster in American Sāmoa

Think about where you spend the most time – are you prepared if a disaster struck there? You may want to increase disaster preparedness at your home or office. Maybe you drive around the island a lot and would like to increase your disaster preparedness in your car. Spend a little time now considering ways to increase your preparedness so that if disaster ever strikes, you will be ready.

The American Red Cross suggests increasing your disaster preparedness by following three simple steps:

1) Get a kit:

There are multiple types of emergency kits, and you should consider whether having one or more emergency kids would make you more prepared to respond to disasters common to the parts of the island where you spend the most time.

- Basic first aid supplies
- Evacuation kit with clothing, medicine, etc.
- Shelter-in-place preparedness kit

Note: Many people have elements of a kit scattered throughout their homes, but do not have an actual kit.

2) Make a plan:

Meet with your family or coworkers to discuss what you will do in case of a disaster event. Consider different scenarios, e.g., Morning versus evening; At home versus at work/school; Hurricane versus earthquake. To create a good emergency plan, you should discuss:

- How will you respond to emergencies at different places you are likely to be?
- Who is responsible to do what?
- What will you do if you are separated?
- Where will you go if you must evacuate?

Important: Review the plan and *practice it*. Everyone will benefit from practice, but especially children (for a home plan) or new hires (for a work plan).

3) Be informed:

- What are the types of emergencies that might happen in your area (**each** area where you spend time)?
- Who are your local authorities are, how to notify them, and how they will notify you during an emergency? (e.g., Sirens? Advisory, watch, warning?)
- What are local evacuation routes?
- Can you take community trainings (CPR, AED, CERT)?

SHARE WHAT YOU LEARN WITH OTHERS
What **you know** could save someone else's life.

Creating an emergency preparedness kit

What should be included in a good kit?

In the immediate aftermath of a disaster, emergency crews may be unable to immediately reach you. It is a good idea to have an emergency preparedness kit available in your home, office, school, and/or vehicle.

Take some time with your family or coworkers to consider what you would want to have in a **quick-grab evacuation kit** if you must leave quickly for safety; or in a **shelter-in-place preparedness kit** if you are trapped or must stay where you are for an extended period for safety.

Health and Wellness Items:

- Water – one gallon per person, per day
- Food – nonperishable, easy-to-prepare, especially requiring no additional cooking or water
- Manual can opener, eating utensils
- First aid kit
- Medications (7-day supply), other medical supplies, and medical paperwork (e.g., medication list and pertinent medical information)
- Sanitation and personal hygiene items

Safety and Logistical Items:

- Flashlight
- Battery powered or hand-crank radio (NOAA Weather Radio, if possible)
- Extra batteries
- Multipurpose tools (e.g., Swiss army knife)
- Extra cash
- Emergency blanket
- Copies of personal documents (e.g., proof of address, deed/lease to home, passports, birth certificates, and insurance policies)
- Family and emergency contact information
- Map(s) of the area

Additional supplies you might want:

- Whistle
- Matches
- Rain gear
- Towels
- Work gloves
- Tools/supplies for securing your home
- Extra clothing, hat and sturdy shoes
- Plastic sheeting
- Duct tape
- Scissors
- Household liquid bleach
- Entertainment items
- Blankets or sleeping bags

The emergency preparedness kit information on these pages is sourced from the Center for Disease Control and Prevention (CDC), the American Red Cross, and Ready.Gov. Please see the Resources section at the end of this booklet to find more resources to help you prepare for disasters.

What are YOUR special needs for an emergency kit at your home or work?

- Glasses /contact lens cases and solution?
- Baby care supplies, i.e., diapers and formula?
- Hearing aids?
- Cell phone / chargers?
- ?
- ?

_____

_____

_____

_____

_____

_____

_____

_____

_____

Disaster and Climate Change Resources

Psychological Recovery from Disasters:

To access support or further information about post-disaster psychological recovery, please visit the American Psychological Association (APA) Psychology Help Center webpage on recovering emotionally from disaster at http://www.apa.org/helpcenter/recovering-disasters.aspx.

Psychology of Climate Change:

To read more about the psychology of climate change, please visit the APA Psychology and Global Climate Change task force website at: http://www.apa.org/science/about/publications/climate-change.aspx.

Disaster Preparedness Information:

Extensive disaster preparedness information is available from

- the Center for Disease Control and Prevention (CDC), http://www.bt.cdc.gov/preparedness/kit/disasters,
- the American Red Cross, http://www.redcross.org/prepare/location/home-family/get-kit, and
- Ready.Gov, www.ready.gov.

Climate Change Information:

The Pacific Regional Integrated Sciences and Assessments program aims to help Pacific Islanders prepare for and manage the risks from climate variability and change. The Pacific RISA is funded by the National Oceanic and Atmospheric Administration (NOAA). Please visit the Pacific RISA at www.PacificRISA.org.